HOW PEOPLE LIVE

ON THE TUNDRA

Nancy Dickmann

BROWN BEAR BOOKS

Published by Brown Bear Books Ltd
4877 N. Circulo Bujia, Tucson, AZ 85718, USA
and
Studio G14, Regent Studios, 1 Thane Villas, London N7 7PH, UK

Text: Nancy Dickmann
Design Manager: Keith Davis
Children's Publisher: Anne O'Daly

Library of Congress Cataloging-in-Publication Data
Names: Dickmann, Nancy, author.
Title: On the tundra / Nancy Dickmann.
Description: Tuscon, AZ : Brown Bear Books Ltd., [2025] | Series: Fast Track: How People Live | Includes bibliographical references and index. | Audience: Ages 5-7 years | Audience: Grades K-1 | Summary: "How people live in the frozen tundra"– Provided by publisher.
Identifiers: LCCN 2023052177 (print) | LCCN 2023052178 (ebook) | ISBN 9781781219744 (library binding) | ISBN 9781781219805 (paperback) | ISBN 9781781219867 (ebook)
Subjects: LCSH: Tundras–Juvenile literature. | Tundra ecology–Juvenile literature. | Tundra animals–Juvenile literature. | Tundra plants–Juvenile literature. | Tundra vehicles–Juvenile literature. | Winter sports–Juvenile literature.
Classification: LCC GB571 .D533 2025 (print) | LCC GB571 (ebook) | DDC 551.45/3-dc23/eng/20231220
LC record available at https://lccn.loc.gov/2023052177
LC ebook record available at https://lccn.loc.gov/2023052178

The photographs in this book are used by permission and through the courtesy of:
Cover: iStock: Evgeny Kharitonov. Interior: Alamy: H. Mark Weldman Photography 15; Public Domain: Ansgar Walk/GNU Free Documentation License, version 1.2 20t; Shutterstock: ActiveLines 10-11bk, 18-19bk, AnnstasAg 8-9bk, 16-17bk, Atosan 20b, Denis Belitsky 18, BNP Design Studio 4-5bk, 12-13bk, Chrispo 6, GraphicsRF.com 6-7bk, 14-15bk, Green shut 3br, Jukka Jantunen 10, N.K. Junky 12, Yongyut Kumsri17, longtaildog 23, Malchevska 24br, maquidoux 21b, Daniel Marian 5, Evgenii Mitroshin 9, 21cr, Mystic Stock Photography 11, Vadim Nefedoff 8, Porcupen 20-21c, Red Monkey 1, 2-3, 20-21bk, 22-23bk, 24, Serge Reshar 19, sirtravelalot 13, Kirill Skorobogatko 14, smutan 1inset, Andrei Stepanov 7, Robert Szymanski 21t, The Nomad Photography 16, Yobab 4.
t-top, b-bottom, l-left, r-right, c-center, bk-background
All other artwork and photography © Brown Bear Books.

Words in **bold** appear in the Words to Know on page 23.

Manufactured in the United States of America
CPSIA compliance information: Batch#AG/5659

Contents

Land of Ice and Snow

The ground is covered in snow.
A cold wind blows. The ground is often frozen.
There are hardly any trees. This is the tundra.

There is tundra high up in mountains.

WOW!

Russia is the world's biggest country. It has the largest area of tundra.

The tundra is a tough place to live. But some animals survive here. They are **adapted** to the cold. People live in the tundra, too.

Weather and Climate

Winters in the tundra are very long.
They last more than half the year.
The days are short. The soil freezes.

In the tundra, the sun is often low in the sky. It doesn't give much warmth.

Summers are short. The days are long. The weather is a little warmer. The top of the soil **thaws**. Plants grow. But they have to be quick. It will be winter again soon.

Tundra Homes

Some tundra people once lived in **igloos**.
These small homes are made of snow.
People cut snow into blocks.
The blocks fit tightly together.

Some tundra people have modern homes. They are made to stay warm.

WOW!

The tent is made for two families. It can take less than a hour to put one up.

Some tundra people in Russia follow **reindeer**. They move every few days. They live in a tent. The tent has a wood frame. It is covered with reindeer **hides**.

Food and Drink

Tundra summers are too short to grow **crops**.
People gather berries and roots.
On the coast, they find seaweed.
But the main food is meat.

Some tundra people hunt reindeer. They eat the meat. They use the hides for clothes.

Tundra towns today have grocery stores. They sell fresh fruit and eggs. They sell snacks and meals. These foods come from far away. They arrive by truck or plane. They cost a lot of money.

Clothes

Tundra clothing must be warm.

People in North America invented the parka.

It's a warm coat with a hood.

The hood was lined with warm fur.

Many people wear parkas today. They are based on **traditional** designs.

WOW!

Traditional parkas were coated with fish oil. This made them waterproof.

Some tundra people wear modern clothes. They buy them at stores. They order online. Many still make clothes from animal hides. They make boots and warm mittens.

Jobs

Many tundra people work on the land.
They hunt or fish. They gather food to eat.
People also make jewelry and art.
They sell it to **tourists**. They can sell it online.

In Russia, people make dolls to sell.

In towns, people work as teachers and nurses. Some work in stores. Sadly, the tundra is under threat. People work to protect it.

Getting Around

Many tundra people live near the coast. They use boats. Some paddle kayaks. These small boats have a covered top. It keeps freezing water out!

Kayaks were made from animal hides. Today many are plastic.

Tundra roads can be covered in snow.
People ride **snowmobiles**.
Some travel on sleds.
Teams of dogs pull the sleds.

WOW!

Inuit sleds don't have nails. The pieces are tied together with leather. This makes them tough and flexible.

Games and Sports

Tundra people must stay strong to hunt and fish. So they play sports! These sports teach balance and control. People learn to kick and throw. They test their skills against each other.

Tourists visit the tundra. They might see the northern lights. These shimmer in the sky.

Russia's Nenets people raise reindeer.
Once a year, they race them.
There are other contests too.
People wrestle. They play tug-of-war.
They try to throw a rope over a pole.

Where in the World?

There areas of tundra all around the world. Here are some of them.

The Inuit people live in northern Canada and Alaska.

Penguins live in the tundra of Antarctica, but people only visit.

Most of Greenland is covered with ice. People live on a small part of the island.

Nenets people live on the tundra in Russia.

There is tundra high in the Alps. People often go there to hike.

Activities

Look on a map to find the tundra that is closest to where you live. How far away is it? How could you get there? Are there any roads?

Each year, tundra people compete in the World Eskimo Indian Olympics. Find out more about some of these traditional sports. Are there any you'd like to try?

The tundra is home to many animals.
Choose one and research it.
How is it adapted to survive in the cold?

Find Out More

Websites

dkfindout.com/uk/history/native-americans/inuit/

ducksters.com/science/ecosystems/tundra_biome.php

kids.britannica.com/kids/article/tundra/399631

Books

Day and Night on the Tundra Mary Boone, Pebble Books 2022

Inuit Katie Lajiness, Big Buddy Books 2017

A Walk in the Tundra Rebecca L. Johnson, Lerner Publications 2021

Words to Know

adapted having a change in body or behavior that helps a plant or animal survive in its habitat

crops plants that people grow to eat

hide the skin of an animal, including its fur

igloo a traditional type of temporary house made from blocks of snow in parts of the Arctic

reindeer a four-legged animal with antlers that lives in the tundra; reindeer are often called caribou

snowmobile a motor vehicle with skis for driving across snow

thaw to warm up enough to no longer be frozen solid

tourist a person who travels to a place on vacation to see the sights

traditional done in the same way for many years

Index